Romero, Libby,
All about crystals and
gems : discovering treas
2021.
33305251097337
sa 06/03/21

DIGGING IN GEOLOGY

All About Crystals AND Gems

Discovering Treasures of the Natural World

Libby Romero

Children's Press®
An Imprint of Scholastic Inc.

Content Consultant
Dr. Wen-lu Zhu
Professor of Geology
Department of Geology
University of Maryland, College Park

Safety note! The activity suggested on pages 42 and 43 of this book should be done with adult supervision. Observe safety and caution at all times. The author and publisher disclaim all liability for any damage, mishap, or injury that may occur from engaging in the activity featured in this book.

Library of Congress Cataloging-in-Publication Data
Names: Romero, Libby, author.
Title: All about crystals and gems: discovering treasures of the natural world / Libby Romero.
Other titles: All about crystals and gems
Description: New York: Children's Press, an imprint of Scholastic Inc., 2021. | Series: A true book | index. | Audience: Ages 8–10. | Audience: Grades 4–6. | Summary: "This book introduces readers to crystals and gems"— Provided by publisher.
Identifiers: LCCN 2020035262 | ISBN 9780531137130 (library binding) | ISBN 9780531137178 (paperback)
Subjects: LCSH: Crystals—Juvenile literature. | Precious stones—Juvenile literature. | Gems—Juvenile literature.
Classification: LCC QE392.2 .R66 2021 | DDC 548—dc23
LC record available at https://lccn.loc.gov/2020035262

Design by Kathleen Petelinsek
Editorial development by Priyanka Lamichhane

No part of this publication may be reproduced in whole or in part, or stored in a retrieval system, or transmitted in any form or by any means, electronic, mechanical, photocopying, recording, or otherwise, without written permission of the publisher. For information regarding permission, write to Scholastic Inc., Attention: Permissions Department, 557 Broadway, New York, NY 10012.

© 2021 Scholastic Inc.

All rights reserved. Published in 2021 by Children's Press, an imprint of Scholastic Inc.
Printed in the U.S.A. 113

SCHOLASTIC, CHILDREN'S PRESS, A TRUE BOOK™, and associated logos are trademarks and/or registered trademarks of Scholastic Inc.

Scholastic Inc., 557 Broadway, New York, NY 10012

1 2 3 4 5 6 7 8 9 10 R 30 29 28 27 26 25 24 23 22 21

Front cover: background: A variety of crystals display bold colors; top: shimmering blue amethyst quartz; top right: columns of aquamarine quartz; bottom: a diamond shows off its shine.

Back cover: This cave in Mexico holds some of the largest crystals ever discovered.

Find the Truth!

Everything you are about to read is true ***except*** for one of the sentences on this page.

Which one is **TRUE**?

T or F Snowflakes are crystals.

T or F All diamonds are valuable gems.

Find the answers in this book.

What's in This Book?

Orpiment (left) and torbernite (right)

The first crystals began forming on Earth 4.4 billion years ago.

The British royal family's Imperial State Crown is covered in gems.

Dig In!

Crystals are fascinating structures. By definition, they are solids whose **atoms** are arranged in a repeating, orderly pattern. Some are formed naturally. Others are human-made in a lab. From **sugar** and **salt** to **snowflakes**, crystals are all around you! Crystals are found in computers and television sets. There are crystals in your body.

A close-up of blue amethyst quartz

You also walk on crystals every time you go to the beach. When certain crystals are cut and polished, they can become the most beautiful and valuable gems in the world. Gems come in many different colors, shapes, and sizes. **Read on to learn more about crystals and gems.**

Earth's center, or inner core, is made of crystallized iron!

Most rocks are made of different types of crystals that grow near each other.

CHAPTER

Understanding Crystals

So, where is a good place to find crystals? In rocks! Rocks are made of naturally occurring, inorganic (or nonliving) substances called **minerals**. There are more than 5,000 different types of minerals on Earth, and most of these minerals occur as crystals. Other crystals form in living things. Sugar crystals, for example, occur naturally in plants.

How Crystals Form

Crystals aren't alive, but they grow. Crystals can form on Earth's surface, underground, and in the atmosphere. A crystal starts with the first building block, a small group of atoms in a specific pattern. As more atoms are added, the crystal gets bigger. Many crystals grow in liquid mixtures, such as saltwater. As the liquid **evaporates**, the atoms link to form one piece of crystal called a seed crystal. The pattern of atoms repeats, and the seed crystal grows. When the liquid is gone, only crystals are left behind.

Salinas Grandes is a large salt flat in Argentina, in South America. The salt flat formed at the site of an ancient saltwater lake.

A close-up of a thick crust of salt at the Salinas Grandes salt flat

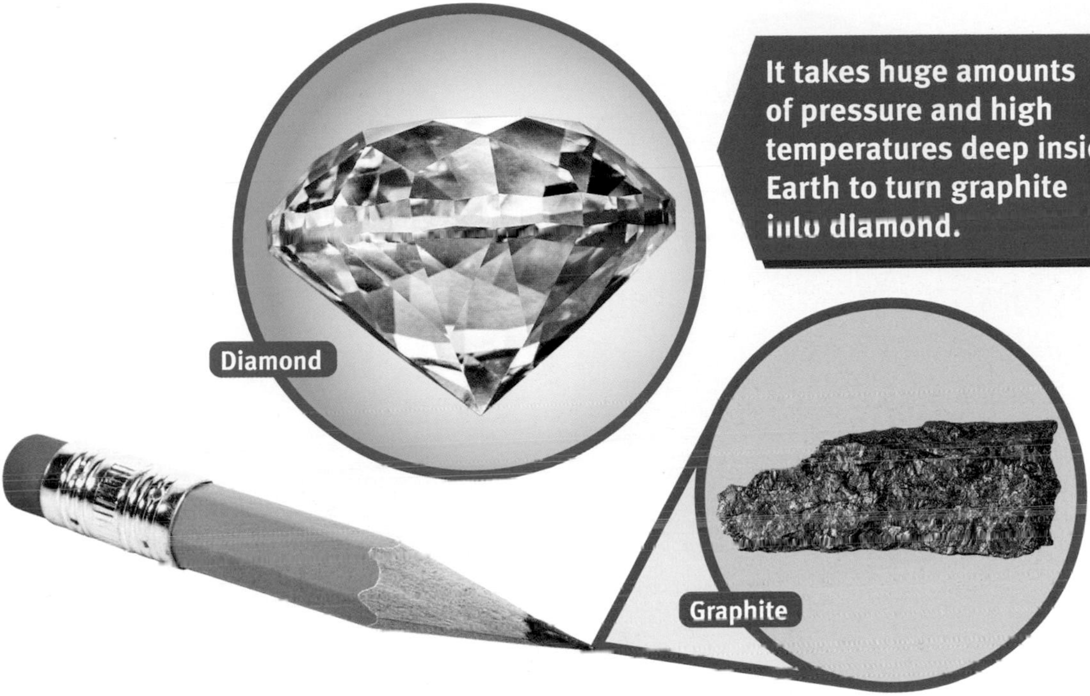

It takes huge amounts of pressure and high temperatures deep inside Earth to turn graphite into diamond.

Diamond

Graphite

Deep within Earth, there are pools of hot, molten rock called magma. Crystals can form here, too. As magma cools, it turns into solid rock. If magma cools slowly, large crystals grow in the rock. If it cools quickly, the crystals are small. When the crystals are exposed to high enough temperatures and pressure, they can transform into different types of crystals. That's how the carbon crystals in graphite—like the lead in your pencil—turn into diamonds!

Garnet

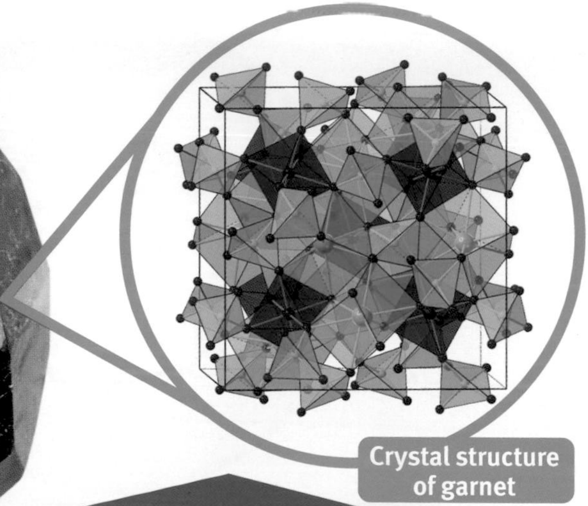

Crystal structure of garnet

A repeating arrangement of atoms within a crystal determines its shape. This is known as a crystal habit. This is how atoms are arranged in garnet.

Crystal Shapes

All crystals have an orderly, internal pattern of atoms that repeats itself again and again. This is known as their **crystal habit**. The crystals of the same kind of mineral always look similar. That's why crystal habit is an important **property** used to identify minerals. In the mineral halite, or rock salt, the crystals are shaped like a box. When garnet forms, the atoms in its crystals have 12 sides—just like soccer balls.

Ice Cold Crystals

Did you know that snowflakes are crystals, too? Ice crystals, to be exact. These winter wonders form when water vapor condenses—changes from liquid to ice—around a speck of pollen or dust in a cloud. This creates a six-sided crystal. As more water vapor condenses onto the crystal, it grows and a pattern emerges. Changes in temperature and humidity as the flake falls determine its shape. Check out these cool ice crystal shapes.

Colors of Quartz

Quartz is one of the most common crystals. All quartz is made from the **elements** silicon and oxygen. Its crystals are shaped like six-sided, pointed prisms. However, all quartz is not the same. Pure quartz (rock crystal) is clear and looks like glass. But sometimes it contains impurities, changing its color. Quartz can be purple, yellow, and brownish gray. It comes in other colors, too!

Each type of quartz is a different color and has a unique name.

Some crystals in this cave in Mexico are about 36 feet (11 meters) long—as long as some school buses!

These crystals may have been growing for more than 500,000 years.

How Big Can Crystals Grow?

Usually, crystals stop growing when magma cools and turns solid or when water evaporates. But what happens if the cooling or evaporation is super slow? Crystals can grow very large! In the year 2000, explorers discovered a cave in Mexico filled with gigantic crystals. How did they get so big? The crystals grew in a huge, humid space filled with mineral-rich waters and a constant temperature near 136 degrees Fahrenheit (58 degrees Celsius).

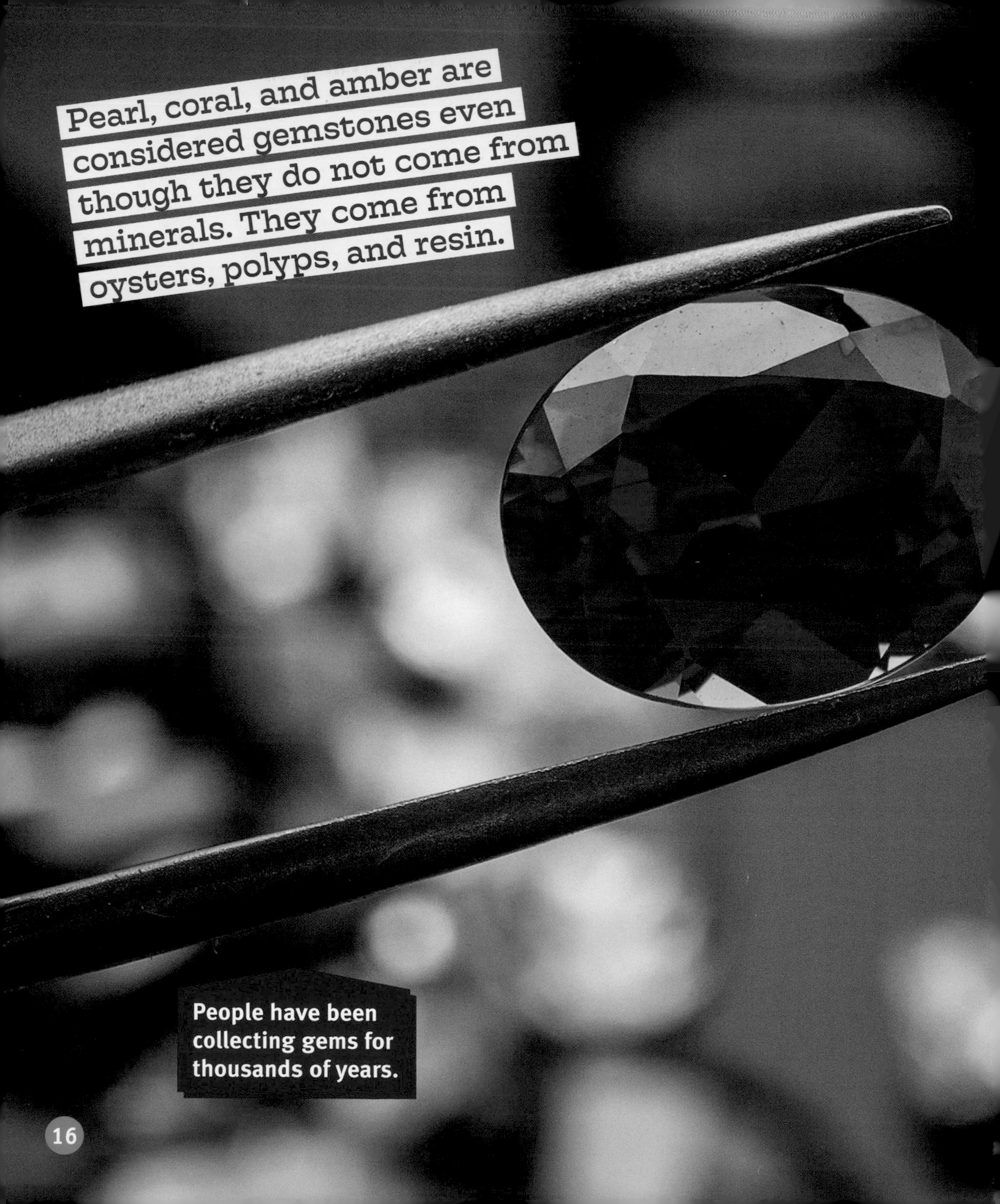

Pearl, coral, and amber are considered gemstones even though they do not come from minerals. They come from oysters, polyps, and resin.

People have been collecting gems for thousands of years.

CHAPTER

2

Gorgeous Gems

Many minerals form beautiful crystals. Fewer than 100 produce crystals that are beautiful enough to be considered gems. Gems are unusually bright, colorful, or transparent materials often used in jewelry. They are relatively rare, and they are extremely strong. Gems are prized for the way they shimmer and shine as light passes through them or reflects off their surfaces. Many people collect gems.

Classifying Gems

Historically, gems have been divided into groups based on how valuable and rare they are. The rarest gemstones are called precious gems. Four gems make up this special category: diamonds, rubies, emeralds, and sapphires. And even within this group, some stones are rarer and more valuable than others. For example, the mineral beryl can form deep green emeralds. But if there is a slight change of elements, it can also produce a super rare variety called red beryl, which is more expensive.

Peridot (left) and garnet (right) are semiprecious gems. They are often used to make jewelry.

All other gems are classified as semiprecious gems. This group of more commonly found gems includes stones like garnet, topaz, aquamarine, opal, and peridot. Although semiprecious gems are still very attractive, they tend to be less expensive than precious gems. This makes them very popular with people who want to buy affordable jewelry.

Rough gemstones can be cut into many different shapes.

The Four C's

A gem's value is based on the four C's: color, clarity, carat, and cut. The most valuable gems have deep, rich colors. Or, in the case of diamonds, no color. Gems with fewer flaws have higher clarity. Carat is the weight of a gem, and cut is its shape or design. In their natural form, gemstones look like ordinary rocks. Gems sparkle when they are cut and polished. Cutting affects how gems reflect (bounce off), refract (bend), and absorb (take in) light.

Light Effects

Light can produce amazing special effects in gems. When light hits a diamond, it reflects a brilliant sparkle and refracts into a rainbow of colors. When it hits the smooth, rounded surface of a gem called tiger's eye, it reflects in a single line. The line moves as you rotate the gem. As light hits a moonstone it appears to glow from inside. And when you look at alexandrite under different types of light, the gem appears to change colors.

Alexandrite's colors can change from bright green in daylight to brown or purple in candlelight.

Famous Gemstones

At 3,106 carats, the Cullinan Diamond from South Africa is the largest diamond ever found. After it was cut, each piece of the diamond was named and numbered. Its largest stones became part of the British royal family's crown jewels. Another gorgeous stone is the deep blue Hope Diamond, once owned by King Louis XIV of France. After several owners went into debt or suffered mysterious deaths, people claimed it was cursed!

Cullinan II Diamond

Cullinan II is one of many jewels that decorate the British royal family's Imperial State Crown.

Birthstones

Birthstones are gems that represent each month of the year. Some people think the stones carry magical powers for those who wear them. Were you born in January? Your birthstone is garnet and could attract friendship, loyalty, and devotion. The modern trend of birthstone jewelry started in Europe in the 18th century.

The BIG Truth

Danger Zone!

Most minerals are crystals. Some are extremely poisonous, or toxic, to humans. Check out these deadly crystals that occur in nature. Learn why it's best for people to avoid them.

Asbestos
Asbestos is a mineral once used to create ceiling tiles, roofing materials, and insulation. Its crystals form thin fibers that break off easily and form dust. Breathing in the dust particles can cause lung cancer.

Cinnabar
People once used this crystal, also known as mercury sulfide, to make paints, jewelry, and even medicines. Not anymore. Touching, eating, or breathing this toxin can cause organ failure, respiratory failure, or death.

Torbernite
The green prism-shaped crystals in torbernite may be pretty, but they're very dangerous. Torbernite is found in granite rocks that contain uranium. It's radioactive! Over time, it releases radon gas that can cause lung cancer.

Galena
Galena, a very common mineral with a brittle crystal structure, is the main source of lead. When you breathe in or ingest its toxic dust, lead builds up in your body to dangerous levels. Even small amounts can poison children.

Orpiment
People used to make poisoned arrows out of orpiment. Its crystals are made of arsenic, one of the most poisonous substances on Earth! Ingesting or touching it can cause lung, skin, or liver cancer.

Diamond is one of the hardest substances on Earth. It's perfect for grinding, drilling, cutting, and polishing other materials.

Diamond-tipped saw blades and drills can cut through rock, concrete, and even steel.

CHAPTER 3

Crystals and Gems at Work

When you think of crystals and gems, do you picture sparkling jewelry or maybe even art? Many people might. But crystals and gems are used for other purposes, too. From saw blades to computers and lasers to solar panels, crystals and gems have many uses. And people have found lots of ways to put them to work.

Computer Chips and Solar Cells

Silicon is an element found in sand. It doesn't conduct, or carry, electricity. But it can if a bit of phosphorous or another element is mixed in. Then its **electrons**, tiny parts of an atom, can move and electricity can flow. Transistors, which make everything from radios to cell phones work, contain silicon crystals. Silicon crystals are also used to create microchips in computers and the energy cells inside solar panels.

Silicon crystals in solar panels absorb sunlight that can be converted into electricity.

Ruby lasers are also found in everyday items like grocery store scanners.

Different types of crystals are used to make laser beams with different colors.

A Gem of a Light

The first laser used a synthetic, or human-made, ruby rod to help direct and concentrate light. When a light bulb wrapped around the ruby rod flashed, the ruby released a beam of pulsing red light: a laser. In 1961, doctors used a pulse from a ruby laser to destroy a tumor in a patient's eye. The operation was quicker and safer than previous methods.

In the Lab

Synthetic gems are made in a lab. They have the same chemical makeup, crystal structure, and properties as natural gemstones. Because of this, they're considered to be real gemstones. Sometimes, the results are so precise that it's hard even for an expert to tell whether a gem is natural or not. Rubies were the first synthetic gems produced. Synthetic gems are also used for jewelry.

Crystals and Gems Through Time

4.4 BILLION YEARS AGO
The first crystals begin forming on Earth.

4000 BCE
Ancient Egyptians begin using gemstones to make jewelry.

MID-1600s
The rare blue-colored Hope Diamond is purchased by a French merchant. It weighs almost 112 carats, making it a very large stone.

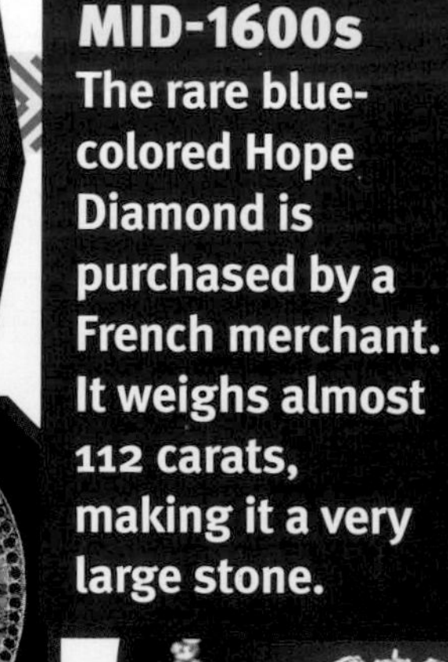

LATE 1800s
The first synthetic gemstones are made in a lab.

Imitation gems are made in a lab, too. These gems can match the color of natural gemstones, but their structure is completely different, so they aren't considered to be real gems. Cubic zirconia, made to look like diamonds, is a good example. Because it is cheaper and also long lasting, like diamonds, cubic zirconia has become a popular alternative. It has also been used to make window material.

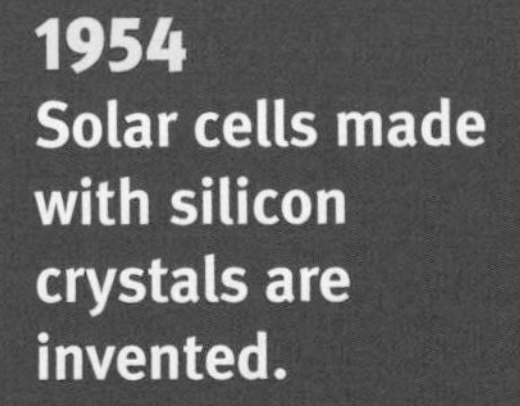

1954
Solar cells made with silicon crystals are invented.

1960
A synthetic ruby is used to build the first laser.

1961
The microchip, powered by a network of tiny silicon crystals, is invented.

TODAY
Extremely tiny nanocrystals help create smaller, faster, more portable electronics.

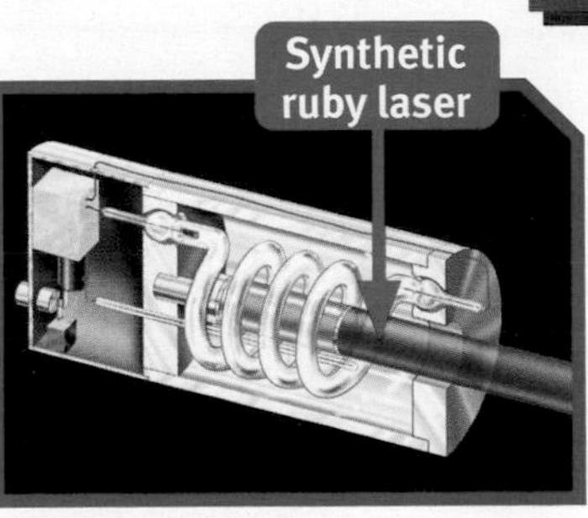

The angled walls in open-pit mines prevent rocks from falling.

Crystals and rough gemstones are extracted from Earth in open-pit mines like this one.

CHAPTER 4

Working with Crystals and Gems

The first people who work with gems are those who get them out of the ground. This is done in different ways: Miners dig in underground mines or open pits to find **deposits** of diamonds and emeralds. And people search in streams to find rubies and sapphires. They pan for them, or swish water and mud through a sieve—a pan filled with holes. Any gemstones left in the pan are collected.

Flaws and fractures make gemstones weaker so they are more easily damaged.

Lapidaries use different techniques to cut and polish gemstones.

Making the Cut

A lapidary is a person skilled at the art of cutting and polishing gemstones. Before the cutting begins, lapidaries study gemstones' crystals. They look for natural flaws or fractures and see how stones reflect and refract light. Then they cut the stones to create flat surfaces called **facets**. After lapidaries cut gems into the desired shapes, they polish them so they shine. Jewelers turn the finished stones into decorative pieces of jewelry.

Tools of the Trade

It takes a skilled craftsperson to create a dazzling gem. It also takes the right tools. Here are just a few of the instruments that lapidaries use.

Caliper: This instrument measures the dimensions of a gemstone.

Dichroscope: This tool is essential for viewing different colors or shades in a gemstone.

Tumbler: Not all gemstones are cut and faceted. Some are just polished in a tumbler until they have a smooth, round finish.

Grit: This finely ground substance is used to cut or polish gems. It can be made of diamonds, aluminum oxide, or many other abrasive materials.

Faceting machine: This machine is used to cut facets on a gemstone.

10X loupe: This handheld magnifier zooms in so lapidaries can see crystals, bubbles, or fractures inside a gemstone. This helps them identify and evaluate a gem. They may also use a microscope.

Drill: Lapidaries use a drill with an extremely fine tip to engrave designs in gems.

Dop: A gem is attached to this metal post with wax to keep it secure during cutting or faceting.

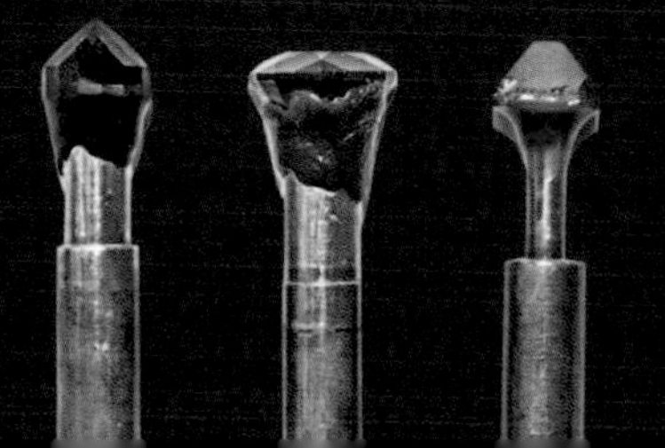

Studying Crystals

Scientists who study the shape and structure of crystals are called crystallographers. Crystallographers use X-rays, computer modeling, and other tools to study the atoms in crystals. They examine how and why crystals grow into certain shapes. Then they arrange crystals into one of seven groups, called **crystal systems**. Their work helps other scientists understand the inner structure of **matter**. Crystallography has led to advances in the fields of genetics, medicine, and chemistry.

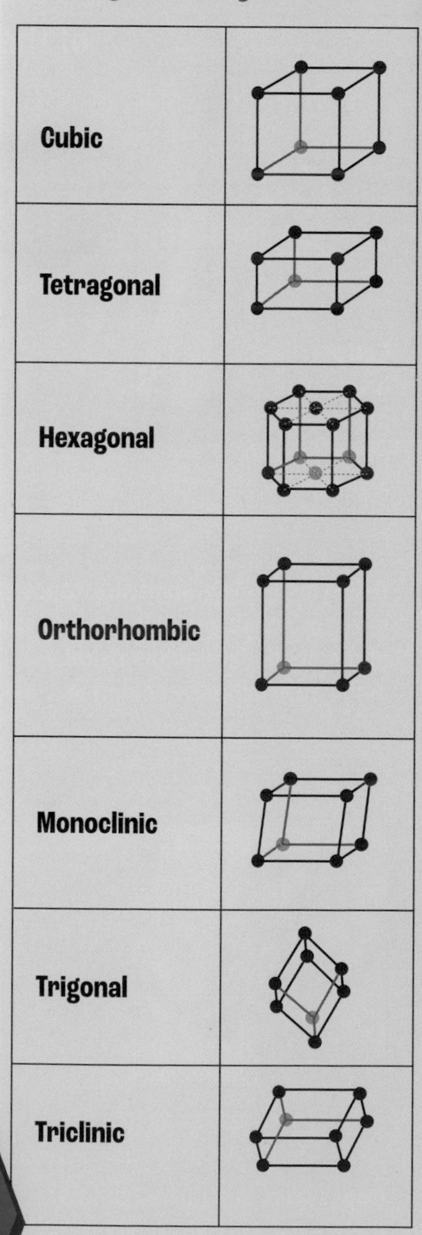

Every type of crystal has a shape, or crystal habit, that fits into one of the seven crystal systems.

Engineers use crystals to create the new technologies that shapes the world we live in.

Crystallography has won more Nobel Prizes than any other scientific field.

Engineering Crystals

Engineers also work with crystals. A chemical engineer can create the synthetic gems found in a jewelry store. An electrical engineer invented the liquid crystals found in computer monitors and TV screens. Other scientists have learned to invent new products like sensors that make cell phones better and faster than before. Crystals and gems are part of our daily lives. They are also the building blocks for the technology that will affect our future.

11 Crystal and Gem Facts

The word *carat* comes from the Greek word *keration*, which means "carob seed."

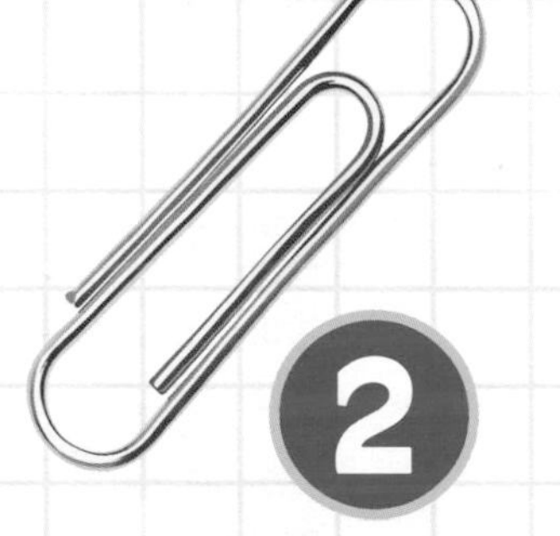

2

Five carats are equal to 1 gram, or about the weight of a small metal paper clip.

3

Diamonds can burn. If the fire is hot enough, they will disappear.

6

The first radios used vibrating crystals to transmit signals.

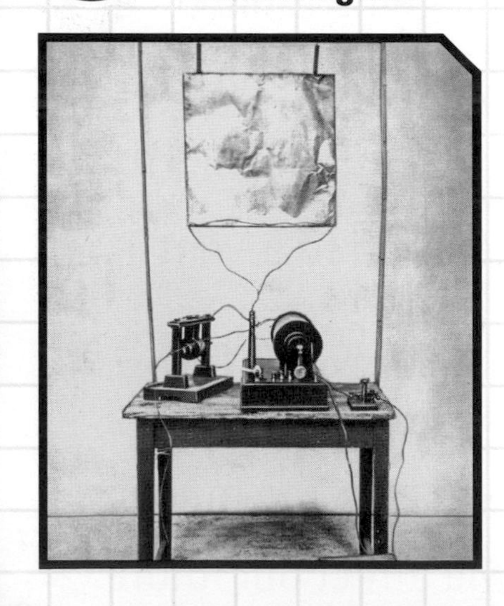

7

When sugar crystals are broken apart, they create light.

8

Tiny calcium carbonate crystals in your inner ear help you keep your balance.

When you squeeze some crystals, like quartz, electricity flows through them.

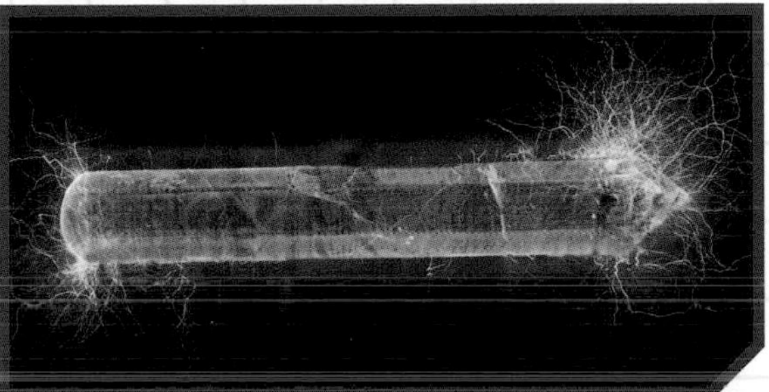

Diamonds can form in outer space and fall to Earth.

9

Sapphires come in many colors, including blue, green, purple, and black.

Turquoise, a semiprecious stone, is sacred to many Indigenous Peoples of North America.

10

Cleopatra loved emeralds. She even had her own emerald mine!

11

Gem Color

To create the most valuable synthetic gems in a lab, chemical engineers must pay particular attention to color. Nothing affects a gem's price more than color. Study the graph and answer the questions that follow.

Analyze It!

1. Which gems are the most expensive?
2. Which are the least expensive?
3. Do you see a trend or pattern in the graph? If so, describe it.
4. Gems in the "ideal color" range are the most valuable. But they are not always the best-selling gems. Why do you think that is? Explain the reasoning behind your answer.

How Color Affects a Gem's Value

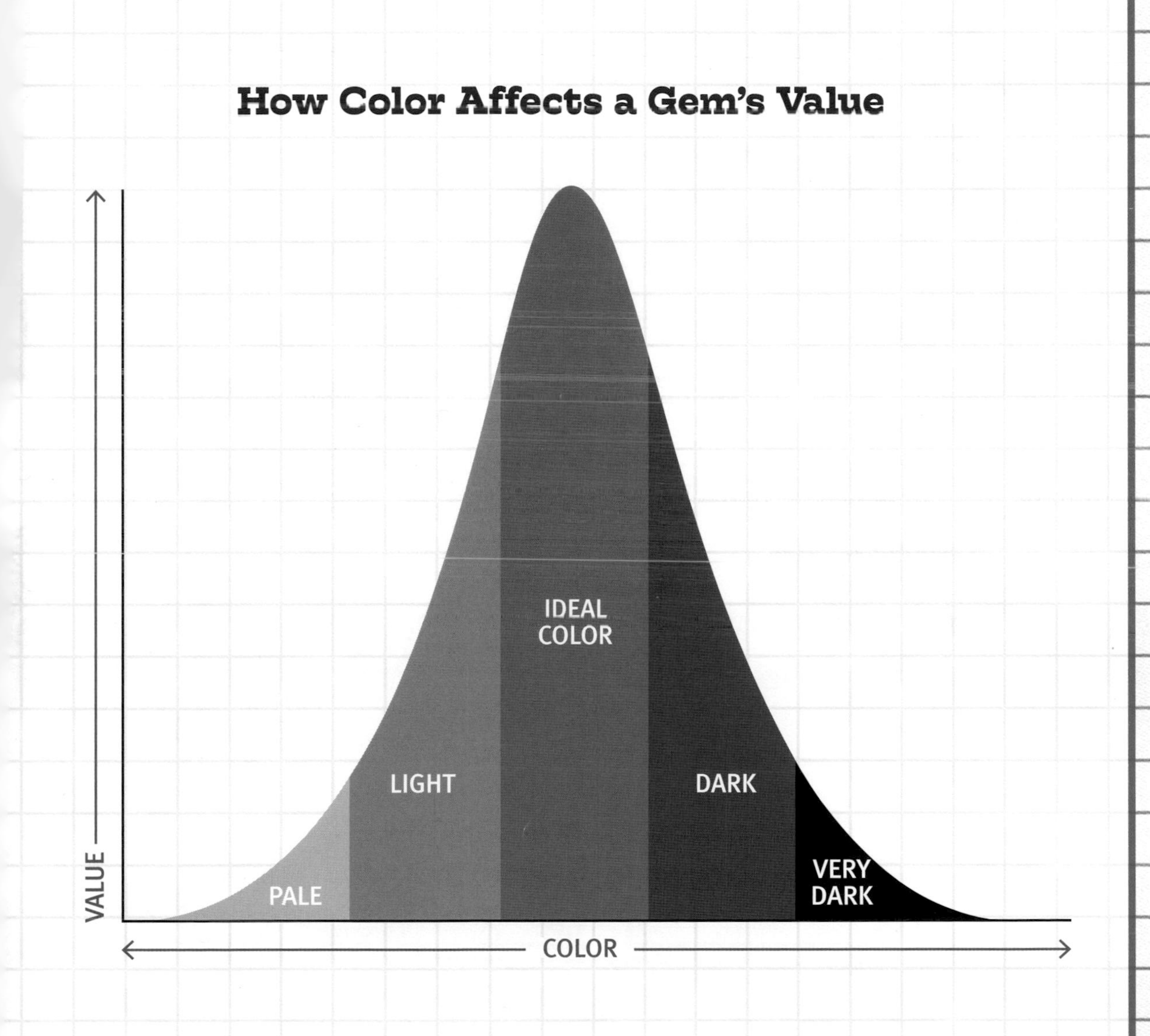

ANSWERS: 1. Those with ideal color, not too light and not too dark. 2. Pale or very dark gems are the least expensive. (Color is less evenly distributed in pale or dark gems.) 3. Yes. The price goes up as the color of the gem gets better and down as the color of the gem gets too dark or light. 4. People like different things. They may like lighter or darker gems, or they may buy them because they're less expensive.

Activity

Grow Crystals at Home!

Rock candy is a fun (and tasty) way to learn how crystals grow. Here is how to make your own! **ADULT SUPERVISION IS NEEDED FOR STEPS 2 AND 3 OF THIS EXPERIMENT.**

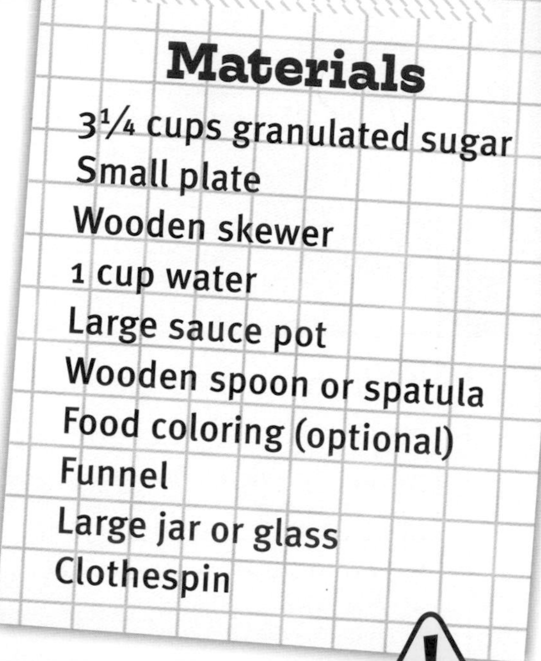

Materials

$3\frac{1}{4}$ cups granulated sugar
Small plate
Wooden skewer
1 cup water
Large sauce pot
Wooden spoon or spatula
Food coloring (optional)
Funnel
Large jar or glass
Clothespin

Directions

1. Pour $\frac{1}{4}$ cup of sugar on the plate. Wet the skewer. Roll it in the sugar on the plate to create seed crystals. Let the stick dry for 30 minutes.

2. Add water to the pot. **With an adult's help**, bring the water to a simmer. Add 1 cup of sugar. Stir until it is dissolved. Repeat with the other 2 cups of sugar. Keep stirring until the water boils.

<u>With an adult's help</u>, add a few drops of food coloring. (To make multiple colors, do this step after completing step 4. You will need multiple jars.)

4 Let the solution cool. Using a funnel, carefully pour it into the jar. Attach the clothespin to the skewer. Insert the skewer into the jar.

5 Set the jar in a safe place and watch it over the next week. What happens to the skewer? How do you think creating seed crystals on the skewers affected your results?

Explain It!

Using what you learned in the book, can you explain what happened and why? If you need help, turn back to pages 10 to 12.

True Statistics

Number of sides on a salt crystal: 6

Temperature at which snow crystals start to form: 32°F (0°C)

Depth of Mexico's Cave of the Crystals: 980 feet (300 m) below the surface

Number of Nobel Prizes connected to crystallography: 29 (as of 2020)

Weight of the Golden Jubilee Diamond, the largest cut and faceted gem in the world: 545.67 carats

Weight of the Bahia Emerald, the largest uncut precious gemstone ever found: 752 pounds (341 kg)

Price of the Pink Star Diamond, the most expensive gemstone ever sold: $71.2 million

Amount of time it takes to grow diamonds in a lab: 6 to 12 weeks

Did you find the truth?

(T) Snowflakes are crystals.

(F) All diamonds are valuable gems.

Resources

Other books in this series:

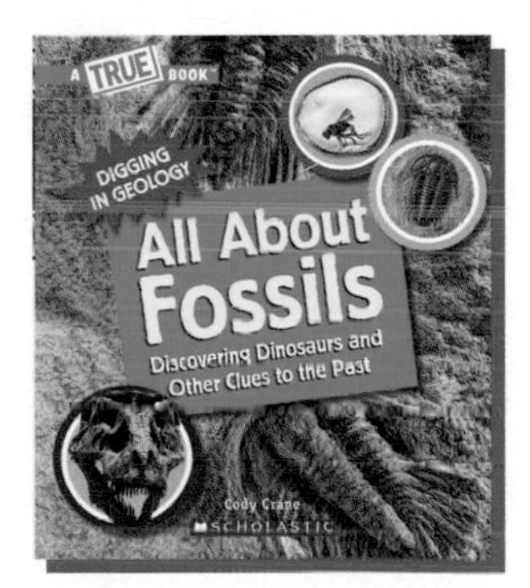

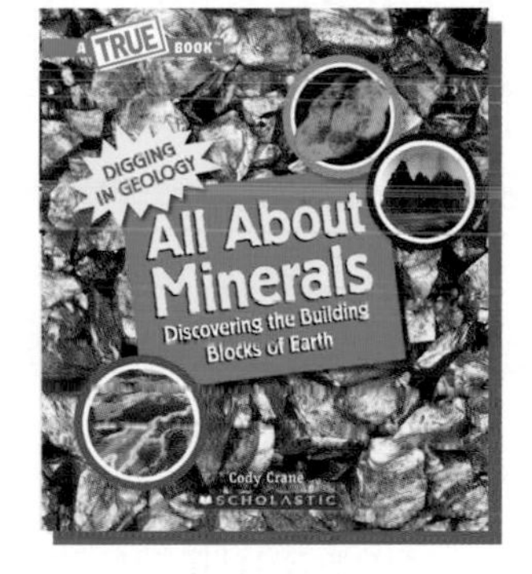

You can also look at:

Green, Dan, and the Smithsonian Institution. *The Rock and Gem Book . . . And Other Treasures of the Natural World*. New York: DK Publishing, 2016.

Polk, Patti. *Collecting Rocks, Gems and Minerals*. Iola, Wisconsin: Krause Publications, 2016.

Symes, R. F. *DK Eyewitness Books: Crystal & Gem*. New York: DK Children, 2014.

Tomecek, Steve. *Everything Rocks and Minerals*. Washington, D.C.: National Geographic Kids Books, 2011.

Glossary

atoms (AT-uhms): the tiniest parts of elements that have all the properties of that element; all the matter in the universe is made up of atoms

crystal habit (KRIS-tuhl HAB-it): the tendency of minerals to repeatedly grow into characteristic shapes

crystals (KRIS-tuhls): substances that form a pattern of many flat surfaces when they become a solid

crystal systems (KRIS-tuhl SIS-tuhms): the seven groups scientists use to classify crystals based on their shape

deposits (di-PAH-zits): natural layers of rock, sand, or minerals

electrons (i-LEK-trahns): tiny particles that move around the nucleus of an atom; electrons carry negative electrical charges

elements (EL-uh-muhnts): substances that cannot be divided up into simpler substances

evaporates (i-VAP-uh-rates): changes liquid into a vapor or gas

facets (FAS-its): flat, polished surfaces of a cut gem

matter (MAT-ur): something that has mass and takes up space, such as a solid, liquid, or gas

minerals (MIN-ur-uhls): solid substances found in earth that do not come from an animal or plant

property (PRAH-pur-tee): a special quality or characteristic of matter, as in the properties of a liquid

Index

Page numbers in **bold** indicate illustrations.

About the Author

Libby Romero was a journalist and teacher before becoming an author. She studied agricultural journalism at the University of Missouri–Columbia (B.S. and B.J.) and received her M.Ed. from Marymount University in Arlington, Virginia. As a child, she read nearly every nonfiction book in her school's library. Now she's added dozens of new books that she's written to the shelves. She lives in Virginia with her husband and two sons.

Photos ©: cover bottom left: Dimitri Otis/Getty Images; back cover: Carsten Peter/Speleoresearch & Films/National Geographic Image Collection; 3: Carsten Peter/Speleoresearch & Films/National Geographic Image Collection; 4 bottom: Björn Wylezich/Dreamstime; 5 bottom: Toby Melville/AFP/Getty Images; 10 all: Libby Romero; 11 top: Dimitri Otis/Getty Images; 12 right: Crystalmaker.com; 13 background: TothGaborGyula/Getty Images; 14 top left: MvH/Getty Images; 15: Carsten Peter/Speleoresearch & Films/National Geographic Image Collection; 18 center bottom: The Natural History Museum, London/Science Source; 19 top: Dani3315/Dreamstime; 19 inset: Winterling/Dreamstime; 20: Phil Robinson/age fotostock; 22: Toby Melville/AFP/Getty Images; 24 left: Vvoevale/Dreamstime; 24 right: Charles D. Winters/Science Source; 25 top: Björn Wylezich/Dreamstime; 29: Wolfgang Kaehler/LightRocket/Getty Images; 30 center left: Bridgeman Art Library/Image Partner/Getty Images; 30 center right: Bridgeman Images; 30 right: PjrRocks/Alamy Images; 31 left: Pasieka/Science Source; 31 center: U.S. Department of Energy; 31 right: Sakkmesterke/Science Source; 34: Frank Bienewald/LightRocket/Getty Images; 35 top left: Mint Images/David Arky/Science Source; 35 top right: Richard Leeney/Dorling Kindersley/Getty Images; 35 center: MonaMakela/Getty Images; 35 bottom right: Hugh Threlfall/Alamy Images; 37: David Wa/Alamy Images; 38–39 background and throughout: billnoll/Getty Images; 38 bottom left: The Granger Collection; 38 bottom right: Fossil & Rock Stock Photos/Alamy Images; 39 top left: Goddard Space Flight Center/NASA; 39 top right: Booth/Garion/Science Source; 39 bottom left: Alexander Potapov/Dreamstime; 39 bottom center: Shizuko Alexander/Getty Images; 39 bottom right: DeAgostini/Getty Images; 42–43 illustrations: Gary LaCoste; 44: Vincent Yu/AP Images.

All other photos © Shutterstock.